CLIMATE SOLUTIONS

ALSO BY PETER BARNES

Pawns: The Plight of the Citizen-Soldier
Who Owns the Sky?
Capitalism 3.0: A Guide to Reclaiming the Commons

CLIMATE SOLUTIONS
A CITIZEN'S GUIDE

Peter Barnes

FOREWORD BY BILL MCKIBBEN

CHELSEA GREEN PUBLISHING
WHITE RIVER JUNCTION, VERMONT

Book design by Peter Holm

Printed in Canada on recycled paper.
First printing, January 2008
10 9 8 7 6 5 4 3 2 1   08 09 10 11

Our Commitment to Green Publishing
Chelsea Green sees publishing as a tool for cultural change and ecological stewardship. We strive to align our book manufacturing practices with our editorial mission and to reduce the impact of our business enterprise in the environment. We print our books and catalogs on chlorine-free recycled paper, using soy-based inks whenever possible. This book may cost slightly more because we use recycled paper, and we hope you'll agree that it's worth it. Chelsea Green is a member of the Green Press Initiative (www.greenpressinitiative.org), a nonprofit coalition of publishers, manufacturers, and authors working to protect the world's endangered forests and conserve natural resources. Climate Solutions was printed on 60-lb. Legacy Offset, an FSC-certified 100-percent postconsumer-waste old-growth-forest-free recycled paper supplied by Webcom.

Library of Congress Cataloging-in-Publication Data

Barnes, Peter.
Climate solutions : a citizen's guide /
Peter Barnes ; foreword by Bill McKibben.
  p. cm.
1. Climatic changes—Government policy.
2. Global environmental change—Government policy.
3. Global warming—Government policy. I. Title.

QC981.8.C5B365 2008
363.738'74--dc22
2007041584

Chelsea Green Publishing Company
Post Ooffice Box 428
White River Junction, VT 05001
(802) 295-6300
www.chelseagreen.com

This book is dedicated to
fellow owners of our one sky.

CONTENTS

Part 3 | Carbon Capping 101

Part 4 | And Finally...

FOREWORD

BY BILL MCKIBBEN

Solving the climate crisis is up to us

In 1992, the first President Bush signed the U.N. Convention on Climate Change, committing the United States to reduce its greenhouse gas emissions to 1990 levels by 2000. Soon after, the U.S. Senate unanimously ratified the Convention.

Since then, our nation has done virtually nothing to meet this commitment. As you read this, our rate of carbon dioxide emissions continues to climb relentlessly.

What will it take for the U.S. to reduce its greenhouse gas emissions rather than increase them? What will it take for us to do this year after year until the Earth's climate stabilizes? That very practical question is what this guide is about.

If you've picked up this guide, you don't need to be told that we face a planetary crisis. You've heard the warnings. You know there's no time to lose. You also know that, although a single citizen can't stop global warming, an army of citizens can.

Fortunately, millions of Americans are now demanding that all levels of government—local, state and federal—take immediate and effective action to cut greenhouse gas emissions. What's more, many thousands are involved in an extraordinary bottom-up policy development process. They're exploring climate solutions and pushing politicians to act. As a result, more than 30 states and 600 cities have adopted policies aimed at cutting carbon emissions. This groundswell has made it a near-certainty that the next President and Congress—the ones who take office in 2009—will finally address the climate crisis at the national level.

Hundreds of proposals are floating about, and many of them aren't very good.

But there's a big problem. Despite countless conferences and think tank reports, there's no consensus on what solutions will actually work. Hundreds of proposals are floating about, and many of them aren't very good. It's quite possible that bad climate policies will be adopted, and that more years will then be lost before real emission reductions occur.

We can't let that happen. That's why you need to read and circulate this citizen's guide. It explains in clear and simple language what different climate policies will do—and, just as importantly, what they won't do. It tells you who's behind the policies, who'd pay for them, and who'd benefit. It demystifies climate policy so that you can play an active role in forming it.

Hundreds of simultaneous rallies have been held across the U.S. as part of the Step It Up campaign to demand government action on global warming. Above, a rally in Charlottesville, Virginia, as part of the first Step It Up National Day of Climate Action on April 14, 2007. Photo by Rose Jenkins. Below, a rally in Centerville, Ohio, as part of the second Step It Up National Day of Climate Action on November 3, 2007. Photo by Scott Knupp.

It's time to choose solutions

In a very real sense, this guide ushers in the next stage of the climate debate. In the first stage, we discussed the problem. In the next stage, we must choose solutions. Should we adopt a carbon tax? A carbon cap? A trading system that allows companies to "offset" their emissions by paying others to plant trees?

> ## We can't wait any longer, and we can't get it wrong.

These are complex questions, but we must come to grips with them. So read this guide and get involved. Join the citizens' army that must solve the climate crisis. We can't wait any longer, and we can't get it wrong.

AUTHOR'S PREFACE

One last chance

In 2006, NASA's top climate scientist warned that we have at most a decade to turn the tide on global warming. After that, James Hansen said, all bets are off. Temperature rises of 3 to 7 degrees Farenheit will "produce a different planet."

If Hansen is right—and most scientists think he is—then every year lost is a year closer to the precipice. In more positive terms, we have one last chance—but only one chance—to save the planet.

This guide is about that last chance. Its two premises are: (1) the climate crisis must be solved now, and (2) popular understanding is a pre-requisite to getting a solution that actually solves the problem.

> We have one last chance—but only one chance—to save the planet.

What's the problem? For many decades, human emissions of greenhouse gases have exceeded the atmosphere's capacity to safely absorb them. We need an economy-wide system to reduce those emissions steadily and surely. If a policy doesn't create such a system, it may be helpful, but it won't be enough.

The atmosphere itself is a commons—a gift of creation to all. It performs many vital planetary functions, including climate maintenance. The trouble is, we humans—and especially we Americans—are disturbing it with our pollution. Even though we know we're doing this, we don't stop. Indeed, we *can't* stop as long as our current system for using the atmosphere persists.

That system—first come, first served, no limits and no prices—is clearly dysfunctional. One alternative is rationing—limit total use and give everyone equal usage rights. Rationing worked during two World Wars, but we're loath to use it again—we prefer market mechanisms to government chits. Such a preference is fine, but it doesn't change the fact that we need an economy-wide system to reduce atmospheric disturbance. The design of that system is what the debate is about.

Here are a few principles that can help us think about that design:

1) The simpler a system is, the more likely it is to work.
2) The fairer a system is, the more likely it is to last.
3) In the future, polluters should pay for the right to pollute.

That third principle is particularly important because, when all is said and done, the debate about system design is a debate about who will pay whom.

> # When all is said and done, the debate is about who will pay whom.

Many large and powerful companies—what I call the legacy industries—are happy with the arrangement in which polluters pay nothing. But pollution has real costs, and if we want to fix the climate crisis, someone must pay them. If polluters don't, the rest of us will. We'll pay them in the form of higher energy prices, and the extra money we pay will reduce our disposable incomes substantially.

About this guide

This guide is intended to help the general reader understand the key measures that must be taken if we are to turn the tide on climate change. A reader who wants additional information can refer to the web sites that are noted at the end.

This guide is also meant to be shared. A free pdf version can be downloaded from www.onthecommons. org, and that version will be updated.

Every author brings certain biases to his work. In my case, having spent three decades in business, I appreciate the dynamism of markets and lean toward systems that steer markets in the right direction. I'm also keenly aware of the fickleness of government policy. In the mid-

1970s and early 1980s, I ran a solar energy business in San Francisco. Thanks to solar tax credits, my company flourished for a few years. But then, President Ronald Reagan abolished the solar tax credits and my company, along with many like it, went bankrupt.

My hope is that the next time the federal government acts, it will irrevocably direct markets away from dirty fuels and toward clean ones. Then, America's indomitable entrepreneurial spirit will eagerly solve the climate crisis.

PART 1 | QUESTIONS

What's the problem?

If we don't understand the problem, it's unlikely we'll be able to fix it. So let's begin by asking, with regard to the climate crisis, what is the problem we need to fix?

Often in public policy, the problem we need to fix isn't immediately obvious. Sometimes we see symptoms without seeing the underlying problem. Other times we see part of the problem but not the whole.

On the surface, global warming appears to be an environmental problem. But deeper down, it's a result of two economic and political failures.

The first of these is a market failure. Humans are dumping ever-rising quantities of carbon dioxide into the atmosphere because there are no limits or prices for doing so. There are, however, huge costs—costs that are shifted to future generations. When people don't pay the full cost of what they're doing, but instead transfer costs to others, economists call this a "market failure." Nicholas Stern, former chief economist at the World Bank, has said that climate change is "the biggest market failure the world has ever seen."

The second cause of global warming is misplaced government priorities. Because polluting corporations are powerful and future generations don't vote, our government not only allows carbon emissions to grow, but subsidizes them in numerous ways. It gives tax breaks to oil companies, spends billions on highways, and devotes a large part of its military budget to defending overseas oil supplies.

> Climate change is the biggest market failure the world has ever seen.

The root causes of climate change are two system failures: a zero price for dumping carbon into the atmosphere, and too many government subsidies for polluting activities and companies. Unless we fix both system failures, we'll never stop climate change. Illustration by Dennis Pacheco.

It's important to recognize that these twin failures permeate our entire economy. They're not problems of the electricity sector, the automobile sector, or the building sector; they're problems of all sectors and must be treated at that level. They distort the behavior of all individuals and businesses. No matter how "responsible" any of us may be, our separate actions can't overcome what these twin failures make most of us do most of the time.

What's required are fixes for both system failures. We need to limit and pay for atmospheric pollution, and we need to shift subsidies from dirty fuels to clean ones. If we don't do both of those things, we won't stop climate change.

What makes good climate policy?

Policies are attempts by government to solve problems. They can be evaluated on three grounds:

1) How effectively do they solve the problem?
2) Whose interests do they serve?
3) What principles do they advance?

Some policies are little more than hot air. They're efforts by politicians to look good without offending their backers.

Many policies tackle only part of a problem. They may achieve small gains, but they don't address the core problem, which continues to get worse.

Some policies are giveaways to private interests. Typically, they're cloaked in public-interest language, but their effect is to enrich a few corporations. Lobbyists work hard to get policies like these.

Many policies don't address the core problem, which continues to get worse.

A few policies genuinely solve big problems, serve the interests of ordinary people, and advance important principles such as fairness and transparency. These are the policies citizens should actively support. Social Security, for example, solves the problem of old-age poverty in a way that benefits all. That same standard should apply to climate policy.

What are the goals?

Any solution to climate change must begin with clear goals. Then, measures must be taken to achieve those goals. It's quite possible that the measures taken will be inadequate, but without clear goals we won't know how we're doing.

Climate goals can be expressed numerically and in terms of system design. The most important numeric goal is to reduce carbon emissions to a level at which the Earth's climate will stabilize.

The most widely accepted scientific study—made by the Intergovernmental Panel on Climate Change—says we must reduce carbon emissions 80 percent by 2050. That works out, on average, to 2 percent of current emissions per year.

It's important to understand what these numbers mean. On the one hand, cutting emissions 2 percent in a year is manageable. It avoids shocking the economy and allows legacy industries to phase out past investments gradually.

On the other hand, cutting emissions 80 percent by mid-century means we have to build an entirely new energy infrastructure. That's no small challenge, but it's one that America can meet. Look what we built in 40 years after World War II. We could do that again, this time with protection of the planet in mind. And in the process, our economy would boom.

In terms of system design, our goal is to build a simple, fair, and market-based system for limiting use of the atmosphere. That too can be done.

Government has four tools for achieving climate goals: taxes, caps, regulations, and investments. It's important to understand the strengths and weaknesses of each. That's what the rest of this guide is about.

Food and gas were rationed during both World Wars, with everyone receiving equal shares. Courtesy Northwestern University Library.

What's fair?

Fairness is one of the most important principles a climate solution should embody. But what exactly is it?

There are many dimensions to fairness. For example, there's interspecies fairness: Are we humans being fair to other species?

There's international fairness: Are we in America, who have emitted more greenhouse gases than any other country, being fair to the rest of the world?

There's inter-generational fairness: Are those living today being fair to their children and grandchildren?

And there's intra-generational fairness: If a policy enriches a small minority, while placing burdens on everyone else, is such a policy fair to our fellow citizens?

> Fairness must be built in from the outset or it won't happen.

The key test for interspecies, international, and inter-generational fairness is: Will this policy reduce U.S. emissions fast enough to prevent planetary catastrophe? If not, we have to try harder.

The key test for intra-generational fairness is: Does this policy share the burdens and gains of curbing climate change more or less equally?

During World War II, the draft applied equally to all males, and rationing meant the same shares for everyone. Fairness wasn't an afterthought; it was built into our policies from the outset.

That should also be true in tackling the climate crisis. It's up to us to make it happen.

Who owns the sky?

If you wanted to dump harmful waste on your neighbor's property, you couldn't do so. Your neighbor would tell you to stop. If you persisted, you could be prosecuted for trespassing.

Alternately, your neighbor could let you dump your waste, but charge you a price for doing so. Either way, you'd have to listen to what your neighbor said.

With the atmosphere, however, things don't work that way. If you dump carbon dioxide into the sky, no one tells you to stop, no one prosecutes you, and no one charges you a penny.

Why is the atmosphere different from your neighbor's property? Because no one effectively owns the atmosphere.

The atmosphere is different from private property—it's a commons belonging to all.

**NO
TRESPASSING
COMMON PROPERTY**

But should, or could, anyone own the sky? And if so, who should it be?

The atmosphere is different from private property—all living beings share it, which makes it a commons. But that doesn't mean its use can't be limited.

As it turns out, it's entirely possible to use property rights and prices to limit use of the atmosphere. But the fact that the atmosphere is a commons means we have to design these tools carefully. We have to make sure that, if the atmosphere is "propertized," the value of those property rights is equitably shared.

It may be helpful to think of the atmosphere as a parking lot for carbon dioxide emissions. In any parking lot, when demand for parking exceeds capacity, we limit use to short time periods and install meters. If the lot is owned by a public entity, the money paid by parkers is used to benefit all.

In the case of our shared atmosphere, we must also limit parking and charge for it. And, the money polluters pay should benefit all.

Who are the players?

Public policies don't arise in a void. They emerge from a political process that's driven by players. As in any competition, it helps to know who the players are. In climate policy, there are four key contenders:

- Legacy industries
- Sunrise industries
- Environmental groups
- Everybody else

The legacy industries are the oil, coal, gas, auto, and electric industries. Their interests aren't identical, but in general, they want to reap maximum return from their past investments and the resources they control. They're happy with the status quo and favor the least demanding changes. Because they've been around a long time—and have lots of money—they enjoy enormous political clout.

The sunrise industries include wind, solar, and some hi-tech companies, plus venture capitalists, investment bankers, and carbon traders. They're comfortable with change and hope to profit from it. In general, they favor subsidies for energy alternatives and lots of carbon trading. But they have less political clout than the legacy industries.

Environmental groups represent, in theory, future generations and the planet as a whole. In reality, they differ substantially in their tactics and alliances. Some prefer market-based policies, others favor government regulation and spending. Some will make deals with polluters, others won't. Overall, their effectiveness in Washington has declined since the 1970s, though lately they've gotten aggressive on climate change.

Fossil fuel industries favor the least demanding changes.

The "everybody else" category is where most of us fit in. We don't own stock in Exxon or a wind company; we do drive cars, pay energy bills, and vote. We want climate solutions that are fair and effective and don't empty our bank accounts. To get that outcome, however, we must make our voices heard. Our disadvantage is that we're poorly organized, but the Internet gives us a boost. At the end of this guide you'll find several ways to connect with climate policy groups through the Internet. Please check them out.

The bottom line in climate policy—as elsewhere—is simple: Unless the public puts pressure on politicians, special interests will rule.

Who pays whom?

Every public policy has winners and losers. Sometimes it's obvious who those are, but more often, it takes some digging to understand how the money flows.

The typical way special interests get money from government is through subsidies and tax breaks. In those cases, all taxpayers pay, and favored companies gain. The wealth transfers can be seen in public budgets.

Subsidies and tax breaks are very much on the table in climate policy debates. In some cases, when they help sunrise industries, they may be good public policy. But climate change presents several opportunities for businesses to enrich themselves at public expense, and citizens must watch carefully.

> In the future, polluters should pay and the public should benefit.

For example, one proposal to cap carbon emissions would give polluting companies free emission permits worth billions of dollars. Other proposals would create a loosely regulated system of carbon offsets that would help traders profit, but add uncertain public benefit.

The big question in climate policy is whether polluters should pay pollutees, or vice versa. If carbon permits are given free to historical polluters, energy prices will rise and we'll all pay more to whoever gets the permits. That wealth transfer—which over time could exceed a trillion dollars—will flow straight from our pockets to the shareholders of private companies. It will be less visible than tax-funded transfers, but a huge shift of wealth nonetheless.

If rewarding polluters is the wrong way to go, the right way is just the reverse. In the future, polluters should pay and the public should benefit. And with good policy that can happen.

Will it last?

The climate crisis won't be solved with the stroke of a pen. Whatever legislation is passed will take decades to implement. During this time, political support for reducing emissions must remain high.

> For climate policy to work, it must be effective for forty years or more.

So it's important to think about political dynamics. What happens when energy prices rise? What happens if a different party comes to power? How likely is it that the initial policies will stay on course?

A look at American history shows that lasting policies

have support from (a) the middle class, or (b) powerful industries. Social Security lasted but the War on Poverty didn't because the former is widely backed by the middle class and the latter wasn't. Subsidies for oil and coal lasted but subsidies in the 1970s for solar and wind energy didn't because the former industries sway more votes in Congress than the latter.

Will the middle class support carbon reductions for 40 years? In large part that depends on who pays whom. If the middle class ends up paying big energy companies, its support will quickly fade. On the other hand, if the middle class gets a fair deal, its support stands a good chance of lasting.

What about technology?

Our technologies got us into this mess, so it's natural to wonder if technology can get us out.

The answer is that technology is part of the solution, but not the whole solution, and not what will drive the solution. Rather, better technologies will emerge when the market and government flaws are fixed.

> Most of the technologies
> we need are already known.

The other truth is, we can't count on a magic techno-fix to rescue us from climate change. Most of the technologies we need to meet our climate goals are already known. The challenge is bringing them to scale. That scaling up can be speeded by public policies, and as that's done, the prices of these technologies will come down.

Here are some other things to understand about technologies.

Fossil fuels are unique

There's no other source of energy that's as concentrated and convenient as fossil fuels. This means we can't simply replace fossil fuels with something else. We also have to use less energy, and use it smarter.

Solar, wind, and tidal power

Solar, wind, and tidal power—like the power of falling water—are free gifts of nature. We've made great strides in harnessing them efficiently. Their chief problem is that they're intermittent and spread out. To take maximum advantage of them, we need a smart electric grid that can move these kinds of power from where they're harvested to where electricity is needed.

Hydrogen

Hydrogen isn't a source of energy—it takes energy to make it. (It has to be extracted from water or fossil fuels.)

Its value is that, once made, it can be stored, transported, and used without emitting greenhouse gases.

> # Hydrogen's usefulness depends on how it's made. If we have to burn carbon to get it, it won't help.

Hydrogen's usefulness as a climate solution depends on how it's made. If it's extracted from water using solar, wind, or tidal power, it will be a boon. If we have to burn carbon to get hydrogen, it won't help much at all.

Nuclear energy

Scientists once thought nuclear power would be "too cheap to meter." It didn't turn out that way. Nowadays, nuclear power is hugely expensive and exists only because of subsidies.

Nuclear energy has another big problem: safety. It's not just that a plant can get out of control (as at Three Mile Island and Chernobyl). It's also that the wastes from nuclear plants are radioactive—and stay that way for thousands of years. On top of that, the same materials that fuel nuclear power plants can be used for bombs. A world with thousands of nuclear power plants would be a dangerous world indeed.

Biofuels

Biofuels are liquid fuels made from plants—ethanol (grain alcohol) and bio-diesel (made from vegetable oil) are best known. They require energy and chemicals to produce, and they emit carbon when burned (though less carbon than fossil fuels). In theory, biofuels can be made from non-edible plants grown on land not suitable for food production, and such new forms of farming should be promoted. But if demand for biofuels rises, it will be hard to stop food farmers from diverting land, water, and other resources to biofuels. That will drive up food prices and raise the question of whether we would we rather drive or eat.

Geo-engineering

Some clever people want to scatter iron filings on the oceans to stimulate growth of phytoplankton. In theory, these oceanic plants would absorb carbon dioxide from the atmosphere. The trouble is, they'd also disrupt marine ecosystems, with unforeseeable consequences.

Other tinkerers would fill the sky with reflective particles that, in theory, would reduce the amount of solar radiation reaching the Earth. This too carries great risks.

The problem with all geo-engineering schemes is that they could take us from the frying pan into the fire. When internal combustion engines were invented, no one imagined they'd disrupt the climate. Similarly, when coolants like Freon were introduced, no one suspected they'd dissolve the Earth's ozone shield. The truth is, we know so little about the Earth's systems that we could easily trigger another disaster by pursuing these strategies on a large scale.

We can do it

There's no doubt that, once Americans make up our minds, we can rise to almost any challenge. During World Wars I and II, we drafted men, sent women into factories, raised taxes, and rationed commodities such as oil and food. In both cases, we not only won the wars but gave our economy huge boosts.

Solving the climate crisis won't require the same degree of mobilization as the two World Wars did, but it will require a new, economy-wide system for limiting our use of the atmosphere, and new priorities for public investment.

Fortunately, if designed right, these climate solutions can be good for our economy. They can spur investment, create jobs, and lift millions out of poverty.

The challenge is to make the necessary fixes and keep them in place for 40 years.

Courtesy Wikimedia Commons.

PART 2 | SOLUTIONS

Four tools

There are four tools government can use to solve the climate crisis: taxes, caps, regulations, and investments. In the end, we'll need a mix, but before we make our brew we need to know the virtues and flaws of each.

Taxes

A carbon tax is a way to charge for dumping carbon dioxide into the atmosphere. It fixes the problem that such pollution is currently free. If the tax is high enough, it discourages businesses and consumers from polluting. It applies to all carbon used in the economy, and it raises revenue for government.

> ## A carbon tax will never be high enough to do the job.

The big problem with a carbon tax is that it has to be very high to decrease pollution sufficiently. When people are addicted to a substance or source of energy, they're willing to pay a lot more before they stop using it. This is as true of fossil fuels as it is of alcohol and tobacco.

A second problem is that, even if a carbon tax discourages individual consumption, a growing population can still generate more pollution.

A third problem is that a carbon tax hurts poor people. In theory this can be mitigated by funneling tax rebates to the poor, but in practice most tax breaks favor the rich.

The ultimate problem, though, is that tax hikes must be approved by politicians, and politicians in America don't like voting for them. When energy prices rise, as they must, our politicians are unlikely to raise them further by adding taxes.

Caps

A carbon cap is a physical limit on the rate of carbon emissions. It fixes the problem that such emissions are currently unlimited.

To implement a cap, the government issues a gradually declining number of emission permits. Once issued, these permits can be traded. (It doesn't matter who emits carbon as long as total emissions decline.) Trading lets markets allocate emission rights among companies that need or want them most. Trading also establishes a variable price for the declining supply of permits.

One of the big questions raised by carbon capping is: Will the government sell emission permits or give them away free? A corollary question is: If the government gives permits away, to whom should it give them? These questions have huge economic significance, since they determine who pays whom to use the atmosphere.

The big advantage of a carbon cap is that it physically limits total emissions. Even if the population grows, total pollution will decline.

The advantage of a cap is that it physically limits emissions.

The chief problem with a cap is that it can lead to higher prices for consumers and windfall profits for certain companies. These problems—and their remedies—are discussed on pages 28–33.

Regulations

Regulations are rules promulgated by government that require businesses to do certain things. They vary from sector to sector and require specific actions by specific dates with fines for non-compliance.

Examples of climate-related regulations are automobile fuel-efficiency standards, renewable-energy portfolio standards (requiring utilities to generate a rising percentage of their electricity with solar or wind power), and efficiency standards for appliances and buildings.

The advantage of regulations is that they make businesses do things they otherwise wouldn't do. The disadvantage is that they're disliked and resisted by the businesses they affect. They also apply only to targeted industries, so a different set has to be adopted for each sector of the economy.

Investments

Public investments have changed the face of our country in the past, and can do so again in the future. In the 19th century, they built canals, railroads, public schools, and land grant colleges. In the 20th century, they built interstate highways, hydropower dams, and the Internet. In the 21st century, they can help build a post-carbon infrastructure.

Investments can be in the form of actual expenditures—for example, grants to cities for mass transit. Or they can be in the form of tax breaks—renewable-energy tax credits, for instance. Their function is to subsidize desirable activities that the market, by itself, doesn't support.

> The challenge with government investments is to make good ones and avoid bad ones, which isn't easy.

The challenge with all government investments is to make good ones and avoid bad ones. This isn't always easy, as government expenditures are often driven by politically powerful companies. These companies reap profits by "rent seeking"—that is, by getting government to throw money their way. Frequently, their "return on investment" is hundreds of times what they spend on lobbyists and campaign contributions.

With that in mind, certain principles should apply to public investments:

- Don't duplicate what private capital can do;

- Phase out subsidies over time.

The rest of this section reviews these four tools in more detail. Then, in the following section, we'll explore the intricacies of carbon capping.

Carbon taxes

Theory

The theory behind a carbon tax is simple. Raising the price of carbon (by adding a tax to it) will discourage businesses and consumers from burning fossil fuels. The higher the tax, the less pollution there'll be.

The main arguments for a carbon tax are:

- It's simple to administer;

- It covers all carbon in the economy;

- Everyone pays it—no one gets a free ride;

- The tax rate is predictable;

- The rules are transparent and easy to understand;

- Revenue can be returned to citizens through tax cuts, or used for public investments.

When carbon is taxed or capped, energy prices will rise. So will the prices of products—including food—that require fossil fuels to be produced and distributed.

The simplicity, transparency and predictability of carbon taxes are often contrasted with the complexity of carbon capping systems.

History

The idea of pollution taxes was first proposed in 1920 by British economist Arthur Pigou. Since then, it has been endorsed by many economists as an ideal tool for fixing market failures. When societal or ecological costs (such as the costs imposed by pollution) are not included in the prices of activities that cause them, government can adjust those prices by adding appropriate taxes.

Pollution taxes have been more popular in Europe than in the U.S. Indeed, gas taxes in the U.S. are used to build more roads, which spurs more gas use, rather than discouraging use of gas.

The closest thing to a carbon tax that was ever introduced in Congress—an energy tax sponsored by President Clinton in 1993—was soundly defeated by oil and coal lobbyists.

Reality

A carbon tax is an economist's dream but a politician's nightmare. The economist imagines that politicians will keep raising the tax until it reduces pollution sufficiently to solve the climate crisis. That assumes heroic behavior by a majority of Congress members for several decades, an assumption not grounded in reality.

> A low carbon tax would create the illusion of action without changing business as usual.

Oil companies have lately decided that a low carbon tax is their favored "solution" to the climate crisis. They reason that such a tax would create the illusion of action without changing business as usual. There'd be no cap on carbon, and usage would continue to rise.

Who's in Favor

- Economists
- American Petroleum Institute

- Some conservatives who'd like carbon taxes to replace income taxes
- Some liberals who'd like carbon taxes to replace payroll taxes

Who's Opposed

- Most politicians

Likely Consequences

- No tax, or a low tax with insufficient reductions in carbon emissions

Concerns

- A tax that is too low will not solve the problem. Indeed, it could worsen the problem by allowing emissions to rise for years to come.
- A tax that is high enough will almost certainly not fly politically.
- A tax without returning money to the people will hurt everyone but the rich.

Carbon caps

Theory

In theory, a descending economy-wide carbon cap is the best way, if not the only way, to guarantee a predetermined decrease in carbon emissions by a predetermined date. That's because it's an absolute limit on emissions rather than just an incentive or regulation.

A carbon cap is more than an incentive; it physically limits carbon dioxide emissions. However, carbon capping is tricky, and can easily be done wrong. The devil is in the details.

A carbon cap would function through the issuance of permits. Each year the number of permits would be reduced. To cut emissions 80 percent in 40 years, the number of permits would be reduced by a yearly average of 2 percent of current emissions.

Because a cap requires permits, it introduces the opportunity to trade those permits. Businesses like this feature because it gives them flexibility in reducing emissions. But it's important to remember that the key to the system is the cap, not the trading.

The key to the system is the cap, not the trading.

Like a carbon tax, a decreasing carbon cap will drive up the price of fossil fuels. As fewer permits become available, their price in the market will rise, and the higher prices will be passed on to consumers. If private companies keep the higher prices, they'll reap windfall profits. If government gets the higher prices, the money can be used for public benefit. If citizens get the higher prices back, they can maintain their current purchasing power.

The main arguments for carbon capping are:

- It physically drives down pollution, which is the only way to ensure sufficient reductions within the time required;

- If done right, it can cover all the carbon in the economy;
- If done right, it can return money to citizens and generate revenue for public investments;
- Businesses prefer a cap to regulations and taxes, and politicians will vote for a cap.

History

The idea of capping and trading pollution permits was developed by economists in the 1960s. It got its first major test with the Clean Air Act of 1990, which applied it to sulfur dioxide emissions from coal-burning power plants. (Such emissions cause acid rain.) The program successfully cut emissions on schedule and is widely considered a success.

In 2005, the European Union applied the sulfur cap-and-trade model to carbon. The resulting scheme is widely considered a failure. It has led to huge windfalls for companies that received free permits, higher prices for everyone else, and no reduction in emissions. The EU is now trying to fix the program.

In 2006, nine northeastern U.S. states formed a Regional Greenhouse Gas Initiative, which adopted a carbon cap for electric utilities. Unlike the EU, most of the U.S. states decided to auction carbon permits rather than give them away to polluters. This will avoid private windfalls and allow the states to invest the revenue in useful ways.

Among the lessons learned from these experiences are:

- Carbon is a special commodity—it is ubiquitous and vital to our economy. It can be capped but not in the same way as minor pollutants.
- Capping carbon has big price impacts and can generate correspondingly big windfalls or revenue.

Reality

Carbon capping can be complex, especially when it involves giving free permits to companies and allowing companies to offset rather than reduce their emissions (see Offsets Aren't Permits, p. 60). If these features are removed, carbon capping becomes simpler, fairer, and more transparent.

If giveaways and offsets are avoided, carbon capping becomes a lot simpler.

Unfortunately, there's intense lobbying by companies to receive free permits, and this could distort the whole system. Politicians in the northeast states stood up to this lobbying, but it's not clear that politicians in other states, or in Congress, will do so.

Several bills in Congress blend free permits with auctions. Others include "safety valves" that allow extra permits to be issued when the price of permits reaches a pre-set level. Such hybrids have many layers of complexity and are far from transparent. In general, they favor historic polluters at the expense of consumers and other businesses.

Who's in Favor

- Carbon caps with free permits are favored by utilities and some environmental groups.
- Carbon caps with price ceilings (also known as "safety valves") are favored by oil companies.
- Carbon caps with auctions are favored by public interest, labor, and some environmental groups.

Who's Opposed

- The Bush Administration
- Some environmentalists oppose carbon trading (though not caps per se) on the grounds that it privatizes pollution and, in some cases, shifts pollution to poor communities.

Likely Consequences

- Several state carbon caps plus, after 2009, a federal one
- Many complexities and inequities that will take time to sort out

Regulations

Theory

The role of regulations is to make businesses do what market forces don't. Even if we use a cap to reduce carbon emissions over the long run, efficiency regulations can be good first steps.

History

In 1975, in the wake of the Arab oil embargo, the federal government began setting Corporate Average Fuel Economy (CAFE) standards for new cars. The goal was to raise average new-car efficiency to 27.5 miles per gallon by 1985. This was achieved, but since then—thanks to a loophole for SUVs—average fuel efficiency has declined.

Some states—notably California—have imposed higher standards. At one point, California required auto makers to sell a small number of zero emission vehicles, but withdrew that requirement after intense pressure from Detroit. Environmental groups such as the Sierra Club have fought to raise national CAFE standards, but the auto industry, which makes more profit on big cars, has held them off.

States have also required electric utilities to promote conservation and renewables. And, through codes, they've boosted energy efficiency in new buildings.

Reality

Regulations are, by their nature, industry-specific and complex, and tend to be resisted by the industries they affect. Typically they improve efficiency, but don't reduce total emissions (because population and energy use grow).

> ## Typically regulations improve efficiency, but don't reduce total emissions.

Automobile efficiency standards have kept billions of tons of CO_2 out of the atmosphere, but haven't reduced our overall consumption of oil. Total U.S. oil consumption has risen 25 percent since 1975.

The same can be said for regulation of utilities and buildings: They've avoided some emissions, but haven't reduced total consumption or emissions. The key question for climate policy is whether mandated efficiency standards can cut total emissions substantially. The record suggests that they're helpful but not sufficient.

Who's in Favor

- Environmental groups such as the Sierra Club and Natural Resources Defense Council

Who's Opposed

- Most regulated industries oppose industry-specific rules and prefer economy-wide policies that don't single them out.

Likely Consequences

- Numerous state and federal regulations
- Improved energy efficiency
- More renewable energy
- Absent other policies, total emissions won't decline much, if at all.

Investments

Theory

Taxes, caps, and regulations are good for reducing carbon emissions. To create a post-carbon infrastructure, however, we also have to build things: more mass transit, smarter electric grids, denser and greener cities. These will require public as well as private investments.

History

The federal government has a long history of spending money on important challenges, including wars, education, dams, highways, and space exploration. Such spending can be financed by bonds as well as taxes.

Reality

The biggest problem with public investment is that it provides no certainty that we'll reduce emissions quickly enough. Investments take time to kick in, and meanwhile, growth in emissions (if not otherwise limited) will continue.

A further problem is political. In theory, the federal government could sink billions of dollars into clean energy infrastructure. But such public investment must overcome a root cause of climate change: the domination of government by legacy industries. Even today, despite everything we know about global warming, Congress spends far more on fossil fuel subsidies than on clean alternatives.

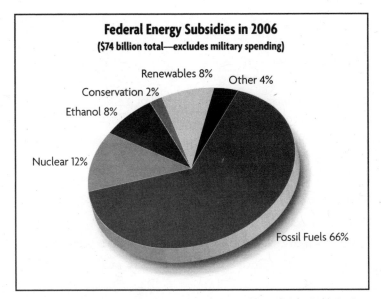

Federal Energy Subsidies in 2006
($74 billion total—excludes military spending)

Renewables 8%
Other 4%
Conservation 2%
Ethanol 8%
Nuclear 12%
Fossil Fuels 66%

Source: Doug Koplow, "Subsidies in the US Energy Sector: Magnitude, Causes, and Options for Reform," exhibit 2, p. 4, www.earthtrack.net

Who's in Favor

Many groups want to shift federal spending on energy. These include:

- Sunrise industries, mayors, environmental groups
- Farmers and agribusinesses who like ethanol subsidies
- Nuclear manufacturers hoping for a revival
- Labor and community groups, such as the Apollo Alliance, seeking job creation

Who's Opposed

- Legacy industries

Likely Consequences

- More subsidies for conservation, wind, and solar after 2009
- More money for ethanol, mass transit, and green job training
- More money for nuclear energy, "clean" coal, carbon sequestration, and other fossil fuel-based technologies

What states can do

Though market failures can best be addressed at the national level, there are many things states and cities can do to fight climate change. Some of these will reduce emissions; others will increase pressure for federal action.

Here are some recent actions taken by states to reduce carbon emissions:

State climate action plans: Typically, the plans set emission reduction goals and propose various policies to meet them. More than half the states have either adopted or are considering such plans.

Regional pacts: In 2005, nine northeastern states formed a regional cap-and-auction system for power plants. Similar pacts are forming in the West and Midwest.

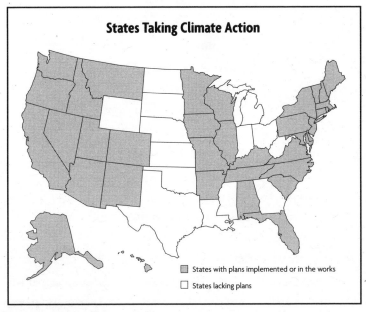

States Taking Climate Action

☐ States with plans implemented or in the works

☐ States lacking plans

States with climate action plans in effect or in planning stages are shaded (as of September 2007). Source: Pew Center on Global Climate Change, pewclimate.org/what_s_being_done/in_the_states/action_plan_map.cfm

Economy-wide reductions: In 2006, California became the first state to adopt economy-wide emission goals. It aims to reduce total emissions 25 percent by 2020 and 80 percent by 2050. It's currently debating the details of a cap and other mechanisms.

Tailpipe emission standards: Fifteen states have pledged to reduce greenhouse gas emissions from new vehicles. A waiver is required from the U.S. Environmental Protection Agency; as of October 2007, it hasn't been granted.

Green building standards: The Leadership in Energy and Environmental Design (LEED) system is a set of standards to improve energy efficiency in buildings. Fourteen states require new state-funded buildings to meet LEED standards.

Renewable portfolio standards: More than 25 states require electric utilities to buy 10 to 25 percent of their energy from renewable sources. This is stimulating demand for wind turbines and large solar arrays.

Clean power purchasing: Connecticut promises to purchase all of its government's electricity from renewable sources by 2050. Maine, Iowa, Illinois, Maryland, New Jersey, New York, Pennsylvania, and Wisconsin also have green-power purchasing plans.

Net metering: Many states require utilities to give credit for solar electricity produced on their customers' homes. In some states, residents who produce more electricity than they use can receive cash back.

Green investment funds: Almost half the states have set

up funds to pay for energy efficiency and renewable energy investments. The funds are collected through small charges on electricity bills.

What cities can do

Urban climate activism has been soaring throughout the country, and in response, nearly 600 cities have pledged to reduce global warming. Here's what some are doing.

Green wheels

Davis, California has created more than 100 miles of bicycle lanes and fully integrated biking as a means of transportation. Most of its downtown is car-free and one out of five residents commutes to work by bike.

New York City Mayor Michael Bloomberg has proposed a daily "congestion fee" for cars that enter downtown Manhattan. Revenue will be used to improve buses and subways. London, Stockholm, and Singapore have similar systems.

Eugene, Oregon requires new developments to include bicycle and pedestrian access.

Seattle, Washington sponsors Rideshare Online, a service that allows commuters to find carpools easily.

Chattanooga, Tennessee builds parking garages on the outskirts of downtown and uses parking revenue to finance electric shuttle buses.

St. Paul, Minnesota has a Neighborhood Energy

Getting people out of cars is a key job for cities.

Consortium that provides energy-efficient cars for shared use.

Austin, Texas' Smart Growth Initiative promotes growth near transit lines and maintains a pedestrian-friendly city center.

Green buildings

Many cities require LEED certification of new and renovated buildings. Others assist building owners in paying for energy-saving upgrades.

In **Scottsdale, Arizona**, more than a third of single-family homes have achieved LEED standards.

Seattle, Washington helps low-income families weatherize their homes.

Portland, Oregon offers low-interest loans and rebates for energy-efficiency improvements.

Radnor Township, Pennsylvania is purchasing wind energy in bulk to meet 60 percent of the town's electricity needs.

Palo Alto, California's municipal utility offers cash rebates to residents and businesses that install solar photovoltaic systems.

Burlington, Vermont subsidizes energy efficiency measures in rental units. The improvements come at little or no cost to apartment owners, and renters save on their monthly energy bills.

San Francisco, California passed a $100 million bond initiative that will finance solar panels, energy efficiency, and wind turbines on public facilities.

Federal preemption

The explosion of state and local climate initiatives is good for the planet but unpopular with large corporations that prefer uniform (and less stringent) national policies. Thus there's mounting interest in Washington in preempting state action.

> Citizens should oppose preemption until strong federal policies are in place.

This could be done prospectively (no future state action could supercede a federal policy) or retroactively (existing state policies could be overridden). Retroactive preemption would unfairly penalize states that acted early. Prospective preemption avoids this, but prevents states from acting more ambitiously than Washington.

Citizens should oppose preemption until strong federal policies are in place.

Cool ideas

Carbon neutrality

Companies, governments, non-profit institutions, and individuals can go "carbon neutral" by purchasing offsets to balance their emissions. The practice has become widespread enough that Oxford University Press named the phrase "carbon neutral" 2006 Word of the Year.

Companies that have gone carbon neutral, or pledged to do so, include Nike, Google, Pepsi, Yahoo!, NewsCorp, and HSBC, one of the largest banks in the world.

Universities that have pledged to achieve carbon neutrality include Cornell, Brown, Johns Hopkins, Middlebury College, and the University of Pennsylvania. Al Gore, Leonardo DiCaprio, Cate Blanchett, George Clooney, and Harrison Ford are among the celebrities who have done this as well.

Legitimate carbon offsets are compatible with, but don't replace, a carbon cap.

At the moment, going carbon neutral is a voluntary and self-enforced choice. However, public policy could require companies and institutions to offset some or all of their emissions in a verified way. This could boost the market for legitimate carbon offsets without undermining a carbon cap.

Electranet

In the future, America will use a mix of large, centralized energy sources (big wind farms and solar arrays) and small, decentralized sources (solar panels on rooftops).

This diversified system will require a new kind of distribution grid.

An Electranet, or smart grid, will allow homeowners and businesses to sell or buy electricity to and from the grid, just as the Internet allows us to download and upload data. It will also enable households to monitor their energy consumption, much as they monitor bank accounts today.

Switching from a one-way grid to an Electranet—and building more long-distance power lines to share energy among regions—will cost billions of dollars, and is a top priority for investment.

New coal moratorium

Despite awareness of global warming, there are more than 150 new coal-burning power plants on the drawing board in the U.S. Ironically, the push to build new coal plants stems in part from anticipation of future emission caps—utilities hope that, if they build new plants before the caps, they'll receive valuable "grandfathered" permits.

Utilities hope that coal-burning plants built before a climate policy is passed will receive valuable "grandfathered" permits.

Many national and local groups are fighting these proposed plants one at a time. They're also targeting banks that finance them. And public figures like Al Gore, John Edwards, and NASA's James Hansen are calling for a nationwide moratorium on new coal plants.

Green collar job training

Several bills in Congress would fund training for workers in energy conservation, renewable energy, and green construction. Such training could provide pathways out of poverty for many young adults and veterans.

Carbon border fees

If the U.S. raises its price of carbon, and countries such as China don't, American manufacturers will be at a competitive disadvantage and U.S. workers could lose jobs.

The 1Sky campaign promotes deep reductions in climate pollution, starting with a freeze on new coal power plants. Logo used with permission of 1Sky.

This problem can be addressed by imposing a carbon border fee on goods from countries that have lower carbon prices than ours. The fee would be based on the amount of carbon required to make the product and the carbon price differential between the U.S. and the exporting country.

> Carbon border fees can protect U.S. manufacturers and encourage other countries to join us in reducing carbon emissions.

Besides protecting U.S. manufacturers, a further benefit of carbon border fees is that they'll spur countries such as China—whose economy depends on exports to America—to constrain their carbon emissions as we reduce ours.

PART 3 | CARBON CAPPING 101

Three varieties of cap-and-trade

One of the tools described in the previous section—carbon capping—deserves special attention, in part because it has several permutations, and in part because it's likely to be adopted in some form.

Carbon capping comes in three varieties: cap-and-giveaway, cap-and-auction, and cap-and-dividend. All start with descending caps. The differences among them lie in who pays whom, and how leaky the caps are.

In cap-and-giveaway, permits are given free to historic polluters. This is called "grandfathering." The more a company polluted in the past, the more permits it gets in the future—not just once, but year after year. As the descending cap raises the price of fossil fuels, everyone

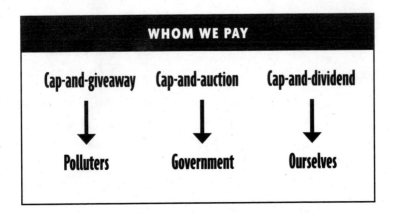

WHOM WE PAY

Cap-and-giveaway → Polluters

Cap-and-auction → Government

Cap-and-dividend → Ourselves

pays more, and the companies that get free permits keep this extra money. Their profits and stock valuations soar, while energy users bear the costs.

Grandfathering would give utilities billions in extra profit for decades.

In Europe, a carbon cap-and-giveaway program handed billions of Euros in windfall profits to a few large utilities. In the U.S., an MIT study estimated that grandfathering permits to American utilities would give them hundreds of billions of dollars in extra profits every year for several decades—a staggering amount of money that would ultimately flow to their shareholders.

In cap-and-auction, permits are sold to polluters, not given away free. Permit revenue is collected by government rather than private corporations. What government does with the money is then up to public officials. It could be used to speed the climate transition, though there are no guarantees.

In cap-and-dividend, permits are also sold, not given away free. However, the revenue doesn't go to the government—it comes back in the form of equal dividends to all of us who pay it. This revenue recycling system is sometimes referred to as a sky trust.

Dividends address the dark side of carbon capping—

the fact that rising carbon prices will take money out of everyone's pockets. According to the Congressional Budget Office, the average U.S. household will pay $1,161 a year in higher energy prices when carbon emissions are reduced 15 percent. As emission reductions increase, so will the cost to households. By recycling higher carbon prices back to households, rebates protect our disposable income while we reduce carbon emissions to safe levels.

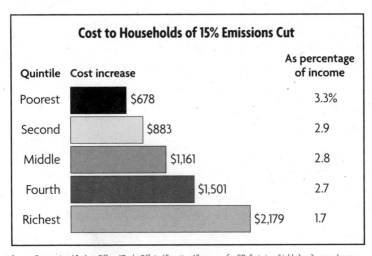

Cost to Households of 15% Emissions Cut

Quintile	Cost increase	As percentage of income
Poorest	$678	3.3%
Second	$883	2.9
Middle	$1,161	2.8
Fourth	$1,501	2.7
Richest	$2,179	1.7

Source: Congressional Budget Office, "Trade-Offs in Allocating Allowances for CO$_2$ Emissions," table 1, p. 2, www.cbo.gov

A seat in Exxon Stadium

It's sometimes said that selling carbon permits will lead to higher prices than giving them away free. After all, why would energy companies raise their prices if they get permits at no cost?

The answer is, businesses set prices by what the market will bear, not by their cost of production. When the supply of carbon permits goes down, companies will charge more for carbon, regardless of what they pay for permits.

If you doubt this, imagine that carbon permits are World Series tickets. If the government gives all World Series tickets to Exxon for free, with no strings attached, will Exxon let people into the stadium for free, or sell tickets for what the market will bear?

Lessons from Europe

To meet its Kyoto Protocol commitments, the European Union set up a carbon capping system in which permits are given free to historical polluters. So far, the results have been dismal. Electricity prices have climbed, coal-burning utilities have reaped windfall profits, and emissions have risen rather than fallen.

There are numerous reasons the EU system hasn't worked. First, because permits are issued to large emitters only, less than half the carbon in the economy is covered. Second, big companies used their political clout to get more permits than they needed (always a danger when companies are given things free). Third, electricity generators who got free permits have raised energy prices and kept the extra income as profit. Fourth, there's no protection for consumers or manufacturers. And fifth, because carbon offsets from outside Europe can be used as substitutes for emission permits within Europe, companies don't actually have to lower their own emissions by the amounts prescribed by the caps.

> Despite the dismal failure of the European trading system, some U.S. senators are promoting a similar cap-and-giveaway system here.

Everyone from Britain's Conservative Party to Germany's Deutsche Bank now says that the system should be radically overhauled. The most important fix is to end the giveaway to large polluters. Says British Conservative spokesman Peter Ainsworth, "The system will not be sorted out until the market is made to work properly by forcing firms to bid for their permits instead of being allowed to lobby government for them free of charge."

Ironically, despite the failure of the European trading system, some U.S. senators are promoting a similar cap-and-giveaway system here. (See Current federal legislation, p. 76.)

KEY LESSONS
- Auction, don't give away, permits.
- Cap all carbon entering the economy.
- Protect consumers and manufacturers.
- Don't count offsets against permits.

Protecting family incomes with equal dividends

A cap-and-dividend system, or sky trust, is a way to reduce carbon dioxide emissions without reducing household income.

The system works by capping carbon, auctioning permits, and rebating the revenue to all residents equally. In this way, it makes everyone pay to burn carbon, but arranges things so that we pay ourselves. As carbon prices rise, so does the money we get back.

> Dividends take politicians off the hook for higher energy prices.

The centerpiece of the system is a trust. (The trust can be run by government or a not-for-profit corporation.) Each year the trust sells a declining number of permits. It then returns the proceeds to residents by wiring money to their bank accounts.

How you're affected depends on what you do. The

Giveaway of the century?

If, through a cap-and-giveaway program, the U.S. Congress hands sizeable chunks of the atmosphere to historic polluters, it will be one of the largest giveaways of a public resource ever. It won't, however, be the first.

Congress has a long history of giving public resources to private corporations. In the 19th century, it gave vast swaths of public land to private railroads. In the 20th century, it gave the public airwaves to private broadcasters. Wisely, it refused to give similar handouts to cell phone companies; instead, it makes them bid for airwave rights at auctions. But giveaways aren't a thing of the past: There's strong pressure now to give the atmosphere to historic polluters for free.

Why do that? Past giveaways were justified on the grounds that the receiving corporations provided some public benefit for the value received. Railroad companies, after all, built railroads. But what will Duke Energy, American Electric Power, and Exelon do for their handouts? Literally, nothing. As the National Commission on Energy Policy, a group consisting partly of energy corporations, has candidly stated, "Giving away emission allowances is like giving away money with no strings attached." The only reason to do it is to buy those companies' political support for carbon capping.

The question this raises, of course, is why a handful of polluting corporations should be granted hundreds of billions of dollars—money that all of us will pay to them in higher prices—just for backing a policy that every American ought to support. If anyone has a right to the economic value of the atmosphere, it's not polluters, but all of us. That's the fundamental reason for auctioning carbon permits.

more energy you use, the more you pay. Since everyone gets the same amount back, you gain if you conserve and lose if you guzzle. Thus, the "winners" are everyone who conserves fossil fuel—plus our children who inherit a stable climate.

The premise of a cap-and-dividend system is that the atmosphere belongs to everyone equally. Its central formula—from each according to their use of the atmosphere, to each in equal share—is fair to poor, middle class, and rich alike. The poor benefit most, however, because they pollute the least.

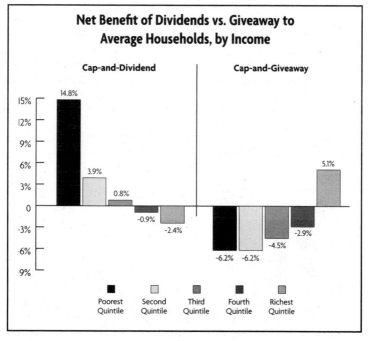

Net Benefit of Dividends vs. Giveaway to Average Households, by Income

Source: James K. Boyce and Matthew Riddle, "Cap and Rebate: How to Curb Global Warming While Protecting the Incomes of American Families," Political Economy Research Institute, Working Paper 150, October 2007, figure 5, p. 35, www.peri. umass.edu

A carbon cap with monthly dividends would be the most popular federal program since Social Security.

From a political perspective, a carbon cap with monthly dividends would be the most popular federal program since Social Security. It would lock in popular support for emission reductions no matter how high fuel prices rise. On top of that, it would take politicians off the hook

At the center of a cap-and-dividend system is a "sky trust," which is like a bank with two windows. One window sells a declining number of carbon permits to fossil-fuel companies; the other pays equal monthly dividends to all Americans. As carbon prices rise, so—automatically—do dividends, thereby protecting household incomes. At the same time, rising carbon prices spur private investment in conservation and clean energy, creating millions of jobs. Illustration by Dennis Pacheco.

for rising prices. If voters complain, politicians can say, "The market sets prices, and you determine by your energy use whether you gain or lose. If you conserve, you come out ahead."

Upstream, downstream

Carbon dioxide doesn't trickle from a few smokestacks; it gushes from virtually everywhere. That makes it hard to cap where it enters the atmosphere. Fortunately, there's a much easier place to cap carbon: where it enters the economy.

Think of carbon as flowing through the economy the way water flows through sprinklers. To reduce the flow of water, you wouldn't plug holes in the sprinklers; you'd turn a valve in the pipe. In like manner, to reduce

Illustration by Dennis Pacheco

the flow of carbon, we can install valves at the relatively few places where carbon enters the economy.

The valves would work like this. All first sellers of carbon-based fuels would be required to buy permits. Each year the number of permits would be lowered. This would physically reduce the amount of carbon flowing through the economy, and eventually into the atmosphere. Economists call this an upstream cap.

An upstream cap would be easy to administer.

An upstream cap would be easy to administer because only a few hundred companies bring fossil fuels into the U.S. During the course of a year, these companies would have to own permits equal to the carbon content of their fuels. Once a year, they'd "true up" and pay a penalty if they don't own enough permits. All carbon would be covered by the cap, and no smokestacks would have to be monitored.

Safety valves and their alternatives

In the context of carbon capping, a "safety valve" is a ceiling price on carbon. When carbon is capped, the price of carbon will rise. With a safety valve, if the price hits a pre-set level, the government issues more permits to keep the price from going higher. The intent is to keep carbon prices predictable and low.

The trouble with a safety valve is that it defeats the purpose of a carbon cap. The issuance of additional permits means, by definition, that the cap will be exceeded. For this reason, oil companies support a safety valve.

There are better ways than safety valves to mitigate the effects of higher prices.

There are better ways to mitigate the effects of higher carbon prices. One is to rebate permit auction revenue to individuals—this protects consumers. Another is to impose border fees on imports from countries with low carbon prices—this protects manufacturers and workers.

Offsets aren't permits

In a simple carbon capping system, companies can trade permits among each other. The permits are issued by government, and there are only a limited number of them.

In recent years, entrepreneurs have come up with a new product: carbon offsets. Sometimes offsets and permits are confused, but they're not the same.

Offsets aren't issued by government, they're not permits to pollute, and there's no limit to how many there can be. Offsets are privately issued certificates that claim to remove carbon from the atmosphere. These claims aren't verified by any government agency.

In theory, purchasing offsets lets you pollute with a clear conscience. Yes, you may have dumped some carbon dioxde into the atmosphere, but your purchase of offsets has presumably reduced emissions somewhere else, so your net contribution to the atmosphere is arguably zero.

Let's say you take an airplane trip. A Web site calculates the tons of CO_2 you emitted. It then sells you offsets, at so many dollars per ton, that purport to withhold an equal amount of CO_2 from the atmosphere.

Offsets imply that we can go about our lives as usual. This isn't true.

There are several problems with such offsets. First, they imply that we can go about our lives as usual; all we need do is "offset" the CO_2 we emit. This isn't true. Until there's an enforceable limit on how much carbon can be dumped into the atmosphere, buying offsets is like playing with Monopoly money—it's a game of pretend.

Second, in many cases, offsets don't actually subtract

CO_2 from the atmosphere. Instead, they pay extra money to private parties for doing things they would or should have done anyway. Offsets bought by Oscar-attending movie stars, for example, went to Waste Management Inc. for cleaning up a methane-emitting landfill that the state had already ordered it to fix.

The danger with offsets isn't just that they may waste people's money. It's that governments will allow them to be substituted for real permits. If that's done, the integrity of any carbon cap would be undermined.

If governments allow offsets to substitute for real permits, they will undermine the integrity of a carbon cap.

Some projects financed with offsets are legitimate. If private buyers want to fund them, that's fine. But whatever emissions are avoided by such projects should be in addition to, not in lieu of, real emission reductions achieved through reducing the number of permits.

In the ideal scenario, there'd be separate markets for permits and offsets, and offsets would add to, rather than diminish, the reductions achieved through permits.

A Satirical Web Site, www.cheatneutral.com

Are you a cheater?
We can help you offset your indiscretions!

Loyal and faithful?
Become an offset project and get paid for not cheating!

What Is Cheat Offsetting?

When you cheat on your partner, you add to the heartbreak, pain, and jealousy in the atmosphere. CheatNeutral offsets your cheating by funding someone else to be faithful and **not** cheat. This neutralizes the pain and unhappy emotion and leaves you with a clear conscience.

Case Study: James and Jo

James and Jo have been together since they met at school. They cheat on each other regularly—James with an ex-girlfriend he can't let go of, and Jo with a man who delivers stationary to her office whose name she doesn't know. To offset their cheating, they fund Chris and Mim through CheatNeutral. In return for payments from CheatNeutral, Chris and Mim promise to remain loyal and faithful to each other so that James and Jo can carry on cheating.

Resolving the carbon price dilemma

By now the reader will appreciate the dilemma we face with regard to carbon prices. From a climate perspective, we want carbon prices to be as high as possible: The higher the carbon price, the less coal we'll burn and the more we'll invest in alternatives such as wind and solar power. But high carbon prices have a cost: They take money from our wallets and move it somewhere else. The higher the carbon price, the lower our disposable incomes. Thus, the dilemma: High carbon prices are good for the planet but bad for households.

HOUSEHOLD INCOMES GO DOWN. SO DOES USE OF CARBON.

CARBON PRICE GOES UP. SO DOES USE OF SOLAR AND WIND POWER.

Illustration by Dennis Pacheco

The key to successful climate policy is to get high carbon prices without hurting households or businesses. That can be accomplished with two complementary tools: dividends to protect households, and border fees to protect businesses. When those tools are added to carbon caps, carbon prices can safely rise to where they need to be.

The key to successful climate policy is to get high carbon prices without hurting households or businesses.

Carbon capping in a nutshell

If done right, a descending economy-wide carbon cap is the single best tool to fight climate change. If done wrong, a cap won't reduce emissions sufficiently and will transfer hundreds of billions of dollars from ordinary Americans to polluting companies and their shareholders.

Doing a cap right means:

- Covering all carbon in the economy
- Selling permits rather than giving them away free
- Paying dividends to residents

CAP AND DIVIDEND IN FOUR EASY STEPS

1. Carbon cap is gradually lowered 80% by 2050.

2. Carbon permits are auctioned.

3. Clean energy becomes competitive.

4. You get an equal share of the permit income.

- No offsets or safety valves
- Protecting businesses with carbon border fees

Doing a cap wrong means:

- Exempting sectors or industries
- Giving polluters free permits
- Putting the burden of higher energy costs on families
- Allowing offsets and safety valves

Carbon Capping EZ Guide		
	RIGHT WAY	**WRONG WAY**
Where to cap	upstream	downstream
How to issue permits	auction	give away
Per capita dividends	yes	no
Offsets	no	yes
Safety valve	no	yes
Carbon border fees	yes	no

PART 4 | AND FINALLY...

An American climate solution

It's time to sum up what we've learned.

First, we need to act quickly—by 2009 at the latest—to fix the market flaw that causes climate change. That means creating a workable and lasting system for limiting our pollution of the atmosphere.

Such a system would reflect the fact that the atmosphere is commons that belongs to everyone. It would cap carbon as it enters the economy, and gradually lower the cap so that, by 2050, emissions are at least 80 percent below the current level.

> We can create a workable and lasting system for limiting our pollution of the atmosphere.

To make sure the cap is airtight, there'd be no safety valves or substituting of offsets for permits.

To prevent stalling or backsliding, the rate at which the cap descends would be set at the outset by Congress, or delegated to an independent trust.

To avoid windfalls to polluters, all permits would be auctioned.

To ensure fairness, sustain middle-class support, and prevent a loss of disposable income, dividends would rise along with energy prices.

To protect U.S. manufacturers and workers, carbon border fees would be added to imports from countries with low carbon prices.

A leak-proof descending carbon cap will have many positive ripple effects. Higher carbon prices will spur private investment in conservation, efficiency, and non-carbon technologies. Utilities will know what kinds of plants to build—and coal won't be on the list. Automakers will know what kinds of cars to build—and they won't be gas guzzlers.

Second, we must also change government priorities. This requires cutting subsidies to fossil fuels and investing in clean energy instead. It also requires higher efficiency standards. The most important measures are:

- A huge investment in mass transit and smart electricity grids;
- Steadily rising efficiency standards for motor vehicles, airplanes, buildings, and appliances;
- Steadily rising renewable energy requirements for electric utilities.

Other helpful policies include:

- Transition assistance to workers, communities, and businesses badly hurt by rising fuel prices.
- Green collar job training.

Of course, legacy industries will resist many of these policies. They'll push for loopholes and giveaways that add emissions and pick our pockets. That's where citizen involvement is critical. Citizens must pressure politicians to hang tough. While it's always tempting to grant concessions to powerful companies, we can't afford to do so this time. The stakes are too high and the margin for error too small.

Fairness

The ultimate test of any climate solution is whether it is fair to future generations. Does it fix the damage this generation has wrought? Does it leave for our children a planet as healthy as the one we inherited?

By this measure, there's no guarantee that any of the measures discussed in this book, singly or together, will be fair. In all likelihood, the historic verdict on late 20th century Americans will be, "They partied, and others paid."

> Will future generations say of us,
> "They partied, and others paid"?

That said, there are degrees of fairness, and it's possible to develop at least an arguable case that one or two sets of climate solutions may be fair to our biological heirs. The key test is whether the solution includes the strongest possible measures to cut greenhouse gas emissions as deeply and as quickly as scientific consensus mandates. Or, putting it in terms a market economy would understand, does the solution push the price of polluting as high as possible? If a solution passes this test, we have a case that we've been fair to future generations. If it doesn't, we've failed them.

In the matter of intra-generational fairness, dictates of science are replaced by precepts of ethics. Because carbon is such a large part of our economy, there's a huge divergence among climate policies in how fairly they distribute gains and losses among living citizens. At the extremes, one solution (cap-and-giveaway) would transfer hundreds of billions of dollars from the bottom 90 percent of Americans to the top 10 percent, while another (cap-and-dividend) would shift a modest amount of wealth in the other direction. Hybrids would fall in between, depending on their mix of giveaways, dividends, and public investments. Where we wind up on this fairness spectrum will be one of the most momentous political issues we face in coming years.

What about China?

Discussions of U.S. climate policy inevitably evoke the question, Why should the U.S. reduce its carbon emissions while China and India continue to increase theirs? This reasonable question is often linked to the fact that, in 2006, China became the largest emitter of CO_2 in the world. The implication of the question is that it's pointless for the U.S. to reduce emissions until China and India reduce theirs. And since that's not happening, the U.S. is off the hook.

There are several responses to this question. First, climate change isn't caused by current emissions alone, but by the buildup of greenhouse gases over decades. Though China now leads in annual emissions, America's cumulative emissions are three times any other country's.

> Global warming isn't caused by current emissions alone, but by the buildup of greenhouse gases over decades.

Second, the U.S. far outpaces China in per-capita CO_2 emissions. Since our population is about one fourth of China's, if the U.S. were to match China's per capita emissions, we'd have to cut our total emissions by about three-fourths.

Third, unlike the U.S., China did adopt the Kyoto Protocol. Under that agreement, industrial nations are required to cut emissions first, with developing nations coming next. That's because the former had two centuries of fossil fuel-burning to get rich, and the latter need some time to catch up.

Chinese leaders are well aware of the threats posed by climate change, but they can't let their country stay poor. They say the U.S. should get its house in order first, and then China will follow. Europeans accept this sequencing and have begun curbing their emissions. Now it's America's turn.

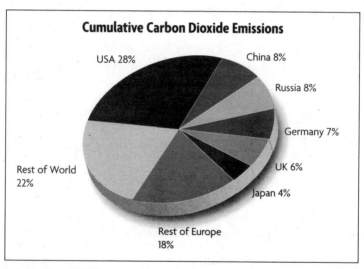

Cumulative Carbon Dioxide Emissions

USA 28%

China 8%

Russia 8%

Germany 7%

UK 6%

Japan 4%

Rest of Europe 18%

Rest of World 22%

Source: James Hansen, "Global Warming: Connecting the Dots from Causes to Solutions," p. 18, www.columbia.edu/~jehl/dots_feb2007.pdf

Beyond Kyoto

No country can stop global warming by itself. Unless all countries move together, every country will beggar its neighbor, and we'll all race to the bottom.

The first attempt at global cooperation—the Kyoto Protocol—was initially supported, then abandoned, by the United States. Kyoto's target—a 7 percent reduction below 1990 emission levels by 2012—was extremely modest, yet few countries will meet it.

The big question now is what will happen after Kyoto expires in 2012. Whereas Kyoto was always seen as a first step, the next treaty must go all the way. It must create a framework to cut emissions 80 percent by mid-century, plus a mechanism for making that happen. And it has to include all major emitters.

The atmosphere is a commons that all people and nations have comparable rights to use.

The core challenge is striking a deal between the U.S. and rapidly industrializing nations such as China, India, and Brazil. This won't be easy. A prerequisite is U.S. leadership, which lately has been lacking. Beyond that, some general principles of equity will have to be agreed.

At the moment, there are several concepts for equity floating about, none of which has U.S. support. Two are:

Contract & Converge is a two-track formula for reducing global carbon emissions equitably. The contract track sets the rate at which global emissions would decline. The convergence track sets the rate at which national emission quotas would approach percapita equality. The year when per capita equality is reached would be negotiated. Both before and after convergence, rich countries could emit more than their share by buying emission rights from poor countries.

The Earth Atmosphere Trust is a plan to curb global warming and end world poverty. It would create a global institution analogous to a sky trust. The trust would auction a declining number of carbon permits and deposit the proceeds in a global fund. A portion of the fund would be returned to everyone on Earth on a per capita basis, or to local community institutions. These payments would be insignificant to the rich but enough to lift the poor out of poverty. The remaining money would pay for renewable energy projects and climate change mitigation.

Both of these concepts rest on the notion that the atmosphere is a commons that all people and nations have comparable rights to use within limits. For any global arrangement to be broadly accepted, it will have to rest on a premise close to that.

Current federal legislation

This summary is accurate as of October 2007. Because legislation changes frequently, the interested reader is advised to check for updates to this page at www. onthecommons.org.

Several bills pending in Congress address the market failure that causes climate change. However, most of them replicate errors of the European trading system: They give free permits to historic polluters, cap carbon downstream rather than as it enters the economy, allow offsets and safety valves, and offer little protection to consumers and businesses. Only one aims to cut emissions 80 percent by 2050, and that one merely authorizes, but doesn't require, a descending emissions cap.

None of the bills is supported by the Bush administration, so their chances of passing before 2009 are slim. That said, they reflect the current state of thinking within Congress, and they'll shape the future debate.

Lieberman-Warner

2020 goal: 10 percent below 2005 level

2050 goal: 70 percent below 2005 level

Initial permit allocation: 76 percent given away free, 24 percent auctioned

Offsets: Yes

Safety valve: Administered by a Fed-like board

Other feature: Most auction revenue goes to fossil fuel companies for research

Bingaman-Specter

2020 goal: 2006 level

2050 goal: Contingent on other countries' efforts

Initial permit allocation: 76 percent given away free, 24 percent auctioned

Offsets: Yes

Safety valve: Starts at $12 a ton

Other feature: Most auction revenue goes to fossil fuel
companies for research

Sanders-Boxer

2020 goal: 1990 level

2050 goal: 80 percent below 1990 level

Initial permit allocation: Authorizes, but doesn't require,
EPA to set a declining cap, auction permits, and
distribute proceeds to individuals, communities, and
companies

Offsets: Not covered

Safety valve: Trigger price linked to a technology index

Other feature: Higher auto and electricity efficiency
standards

Legacy industries on climate policy

The following sections contrast what legacy industries
and religious leaders say about climate policy. They
illustrate the differences between organizations driven
by profit maximization and those driven by moral
responsibility.

American Petroleum Institute

Allowance allocations systems present deep issues of
equity and potential for unfair apportionment among

and within sectors. Equitable and transparent treatment of emissions from different sectors is vital.

The safety valve concept should be considered in any climate proposal.

Exxon

"It's important to get a uniform and predictable cost for carbon across the economy and then let markets pick the technologies that can deliver reductions. The policy that gives you the clearest number is the carbon tax. But there are other options, such as an upstream cap-and-trade system with a safety valve or ceiling price." —Ken Cohen, Exxon vice-president for public affairs

Alliance of Automobile Manufacturers

Any policy to reduce greenhouse gas emissions must focus on all sectors of the economy. CAFE alone cannot achieve this goal; it is a one-dimensional and incomplete program. A broad policy must start with fuel producers and end with fuel users. The further upstream a cap is, the more efficient and effective it is.

Edison Electric Institute

A near 100 percent (free) allocation—with a small percent reserved for auctions—would be recommended. We also recommend that Congress, not an administrative agency, allocate allowances.

EEI supports the robust use of a broad range of domestic and international greenhouse gas offsets.

Faith principles on climate policy

National Council of Churches

We must acknowledge that global warming's impact falls most heavily on poor and vulnerable populations.

We must require that legislation:

- Focus on a fair and equitable distribution of total benefits and costs among people, communities, and nations;

- Support energy sources that are renewable, clean, and avoid destruction of God's creation.

Evangelical Call to Action

When God made humanity, he commissioned us to exercise stewardship over the Earth and its creatures. Climate change is the latest evidence of our failure to exercise proper stewardship.

The most important immediate step that can be taken is to pass and implement national legislation requiring sufficient economy-wide reductions in carbon dioxide emissions through cost-effective, market-based mechanisms.

As a society and as individuals, we must help the poor adapt to the significant harm that global warming will cause.

U.S. Conference of Catholic Bishops

The atmosphere that supports life on Earth is a God-given gift, one we must respect and protect. It unites us as one human family.

Affluent nations such as our own have to acknowledge the impact of voracious consumerism.

The common good requires solidarity with the poor who are often without the resources to face many problems, including the impacts of climate change.

Coalition on the Environment and Jewish Life

Minimizing climate change requires us to learn how to live within the ecological limits of the Earth so we do not compromise the ecological or economic security of those who come after us.

Humankind has a solemn obligation to protect the integrity of ecological systems so that their diverse constituent species, including humans, can thrive.

Nations' responsibility for reducing greenhouse gas emissions should correlate to their contribution to the problem. The United States has built an economy highly dependent on fossil fuel use that has affected the entire globe and must therefore reduce greenhouse gas emissions in a manner that accounts for its share of the problem.

ADDITIONAL RESOURCES

Internet resources

Carbon taxes

Carbon Tax Center, www.carbontax.org

Resources for the future paper, "A Carbon Tax in Time, Saves Nine," www.rff.org/rff/News/Coverage/2005/June/ACarbonTaxInTimeSavesNine.cfm

John Dingell, "The Power in the Carbon Tax," www.washingtonpost.com/wpdyn/content/article/2007/08/01/AR2007080102051.html

Investments and subsidies

Earth Track, www.earthtrack.net/earthtrack/index.asp?catid=73

Carbon caps

U.S. EPA, www.epa.gov/airmarkets/captrade/index.html

Resources for the Future, www.weathervane.rff.org/policy_design/cap_and_trade.cfm

Resources for the Future, "Carbon Emission Trading Costs and Allowance Allocations: Evaluating the Options," www.rff.org/Documents/RFF-Resources-145-c02emmis.pdf

What states can do

Union of Concerned Scientists, www.climatechoices.
   org

Center for Climate Strategies, www.climatestrategies.
   us

Vote Solar, www.votesolar.org

Pew Climate Center, www.pewclimate.org

Regional Greenhouse Gas Initiative, www.rggi.org

Western Climate Initiative, www.westernclimate
   initiative.org

California Climate Change Portal, www.climatechange.
   ca.gov

What cities can do

Local Governments for Sustainability (ICLEI), www.
   iclei.org

Sierra Club Cool Cities Campaign, www.coolcities.us

US Mayors Climate Protection Agreement, www.
   seattle.gov/mayor/climate/

Climate Protection Manual, www.climatemanual.org/
   cities/index.htm

Green Guide Top Cities, www.thegreenguide.
   com/doc/113/top10cities

Lessons from Europe

Financial Times, "Big profits predicted for generators,"
   www.ft.com/cms/s/ed6f3c9c056411dcb151000b5df1
   0621,dwp_uuid=3c093daaedc111db8584000b5df106
   21.html

Guardian (UK), "Smoke alarm: EU shows carbon trading is not cutting emissions," business.guardian.co.uk/story/0,,2048733,00.html

Climatepolicy.com research paper, "Auctioning of EU ETS phase II allowances: how and why?" www.electricitypolicy.org.uk/pubs/tsec/hepburn.pdf

Jörg Haas and Peter Barnes, "Who gets the windfall from carbon trading, or why the European emissions trading system should be transformed into a sky trust," www.boell.de/downloads/oeko/EU_Sky_trust_final.pdf

Cap-and-dividend

ClimateDividends.org, "What Are Climate Dividends?"

Jonathan Alter, "A Clear Blue-Sky Idea," www.newsweek.com/id/33978

Robert Reich, "Carbon Auction's Your Winner," marketplace.publicradio.org/shows/2007/06/20/AM200706202.html

Corporation for Enterprise Development, "Sky trust proposal," www.cfed.org/focus.m?parentid=34&siteid=47&id=93

Peter Barnes and Rafe Pomerance, "Sky Trust: How to Fight Global Warming," www.ourfuture.org/projects/next_agenda/ch10.cfm

Peter Barnes and Marc Breslow, "Pie in the Sky? The Battle for Atmospheric Scarcity Rent," www.peri.umass.edu/fileadmin/pdf/working_papers/working_papers_1-50/WP13.pdf

Congressional Budget Office, "Trade-Offs in Allocating Allowances for CO_2 Emissions," www.cbo.gov/ftpdocs/80xx/doc8027/0425Cap_Trade.pdf

Fran Korten, "Don't Give Away the Sky," www.yesmagazine.org/article.asp?ID=1917

Giveaway of the century?

National Commission on Energy Policy, "Allocating Allowances in a Greenhouse Gas Trading System," www.energycommission.org/site/page.php?report=32

MIT Joint Program on the Science and Policy of Global Change, "Assessment of U.S. Cap-and-Trade Proposals," web.mit.edu/globalchange/www/MITJPSPGC_ Rpt146.pdf

Upstream, downstream

Robert Repetto, "National Climate Policy: Choosing the Right Architecture," www.climateactionproject.com/docs/Repetto.pdf

Andrew Keeler, "Designing a Carbon Dioxide Trading System: The Advantages of Upstream Regulation," www.cpcinc.org/assets/library/9_7keelerjul0.pdf

Tim Hargrave, "U.S. Carbon Emissions Trading: Description of an Upstream Approach," www.ccap.org/pdf/upstpub.pdf

Offsets

BusinessWeek, "Another inconvenient truth," www.
businessweek.com/magazine/content/07_13/
b4027057.htm?chan=search

Financial Times, "Beware the carbon offsetting
cowboys," www.ft.com/cms/s/dcdefef6f35011db984
5000b5df10621,dwp_uuid=3c093daaedc111db858400
0b5df10621.html

Carbon Trade Watch, "The Carbon Neutral Myth,"
www.carbontradewatch.org/pubs/carbon_neutral_
myth.pdf

Joe Romm, "Romm's rules of carbon offsets," gristmill.
grist.org/story/2007/6/29/1170/23713

Legacy industries

American Petroleum Institute, www.api.org

Alliance of Automobile Manufacturers, www.
autoalliance.org

Edison Electric Institute, www.eei.org

Global solutions

Contract and Converge, www.gci.org.uk

Earth Atmosphere Trust, www.earthinc.org/earth_
atmospheric_trust.php

Paul Baer and Tom Athanasiou, "Frameworks &
Proposals: A Brief, Adequacy and Equity-Based
Evaluation of Some Prominent Climate Policy
Frameworks and Proposals," www.boell.de/
downloads/global/global_issue_paper30.pdf

Groups to connect with

Canadian environmental orgainzations
Environment Canada, www.ec.gc.ca
Canadian Association for Renewable Energies, www.
    renewables.ca
Canadian Environmental Network, www.cen-rce.org
The David Suzuki Foundation, www.davidsuzuki.org

Community organizations
Ella Baker Center for Human Rights, www.
    ellabakercenter.org

Faith groups
Interfaith Power and Light, www.
    interfaithpowerandlight.org
Coalition on the Environment and Jewish Life, www.
    coejl.org
Links to Faith and Environment Groups, www.
    sierraclub.org/partnerships/faith/websites.asp

Global networks
EcoEquity, www.ecoequity.org
Global Commons Institute, www.gci.org.uk
FEASTA, www.feasta.org

Labor organizations
Apollo Alliance, www.apolloalliance.org
BlueGreen Alliance, www.bluegreenalliance.org

State and regional grassroots groups

Clean Air Cool Planet, www.cleanaircoolplanet.org

U.S. environmental organizations

Sierra Club, www.sierraclub.org

Natural Resources Defense Council, www.nrdc.org

Greenpeace, www.greenpeace.org

Union of Concerned Scientists, www.ucsusa.org

U.S. grassroots groups

MoveOn.org, www.moveon.org

Alliance for Climate Protection, www.climateprotect.
   org

Step It Up, www.stepitup2007.org

U.S. Public Interest Group, www.uspirg.org

1SKY, www.1Sky.org

Youth and college groups

Focus the Nation, www.focusthenation.org

Campus Climate Challenge, www.climatechallenge.org

It's Getting Hot in Here, www.itsgettinghotinhere.org

GLOSSARY

Allocation: The way in which carbon permits are distributed.

Allowance: A right to emit or burn a specified amount of carbon. The same thing as a permit.

Biofuels: Liquid fuels made from plants.

Cap-and-auction: A carbon capping system in which a declining number of permits are sold to fossil fuel companies and the revenue goes to government.

Cap-and-giveaway: A carbon capping system in which permits are given free to polluting companies.

Cap-and-dividend: A carbon capping system in which revenue from permit sales is returned to individuals via equal dividends.

Carbon credit: Same as an allowance or permit. Can refer to offsets as well.

Carbon neutrality: Balancing carbon emissions with offsets, so that net emissions are zero.

Carbon sequestration: Using technology or natural processes to capture CO_2 from the atmosphere and store it for long periods of time.

Carbon trading: Buying and selling permits and offsets in open markets.

Commons: Resources that are collectively owned and enjoyed by a community. Examples include the atmosphere, ecosystems, the Internet, and the oceans.

Congestion pricing: A fee for using crowded streets or highways.

Dividend A periodic distribution of revenue from carbon permit sales.

Downstream carbon users: End users of fossil fuels. Regulating downstream users is much harder than regulating upstream sellers.

Grandfathering: Giving carbon permits free to historic polluters.

Greenhouse gas: A gas that traps the planet's outgoing heat, thereby causing temperatures to rise. Carbon dioxide and methane are prime examples.

Green or clean energy: Energy that produces no CO_2 and has little or no negative impact on the environment. Examples include solar and wind energy.

Leakage: Holes in a carbon regulatory system that let companies avoid reducing emissions. Offsets and safety valves will result in leakage.

Market failure: When the true cost of something is much larger than the price people pay.

Offsets: Privately sold certificates that claim to remove CO_2 from the atmosphere.

Permit: A government-issued right to emit CO_2.

Renewable portfolio standard: A policy that requires electricity providers to generate a percentage of their power from renewable sources.

Safety valve: A requirement that more permits be issued when the price reaches a pre-set level.

Sky trust: A system for capping CO_2 emissions, auctioning permits, and rebating the revenue to individuals.

Tax shifting: A revenue-neutral shift from taxing income to taxing carbon.

Upstream sellers: Companies that bring burnable carbon into the economy.

Windfall profit: A sudden and unearned profit. If energy companies are given free carbon permits, they'll collect windfall profits at the expense of energy users.

ABOUT THE AUTHOR

Peter Barnes is an entrepreneur and writer who has founded and led several successful businesses. He is presently a senior fellow at the Tomales Bay Institute in Point Reyes Station, California.

Barnes grew up in New York City and earned a B.A. in history from Harvard and an M.A. in government from Georgetown. He began his career as a reporter on *The Lowell Sun* (Massachusetts), and was subsequently a Washington correspondent for *Newsweek* and west coast correspondent for *The New Republic*.

In 1976, he cofounded a solar energy company in San Francisco and, in 1983, he cofounded Working Assets Money Fund. He subsequently served as president of Working Assets Long Distance.

He has served on numerous boards of directors, including the National Cooperative Bank, the California State Assistance Fund for Energy, the California Solar Industry Association, Businesses for Social Responsibility, the Rainbow Workers Cooperative, Redefining Progress, the Family Violence Prevention Fund, Public Media Center, TV-Turnoff Network, the Noise Pollution Clearinghouse, Greenpeace International, the California Tax Reform Association, and the Center for Economic and Policy Research.

His books include *Pawns: The Plight of the Citizen-Soldier* (Knopf, 1972), *The People's Land* (Rodale, 1975), *Who Owns the Sky?* (Island Press, 2001), and *Capitalism*

3.0: A Guide to Reclaiming the Commons (Berrett-Koehler, 2006). His articles have appeared in *The Economist, The New York Times, The Washington Post* and elsewhere.

In 1997, he founded the Mesa Refuge, a writers' retreat in northern California. He has two sons, Zachary and Eli; a partner, Cornelia Durrant; and a dog, Smokey.

ACKNOWLEDGMENTS

The author wishes to thank the following individuals and organizations:

Individuals: Grant Abert, Marcellus Andrews, Jason Barbose, Harriet Barlow, James Boyce, Wes Boyd, Chuck Collins, Peter Dorman, Cornelia Durrant, David Fenton, Lenny Goldberg, KC Golden, Ann Hancock, Andrew Hoerner, Jon Isham, Van Jones, Doug Koplow, George Lakoff, Matt Lappé, Kathleen Maloney, Bill McKibben, Ana Micka, David Morris, Robert Perkowitz, Julie Ristau, Jonathan Rowe, Mike Sandler, Rob Sargent, Betsy Taylor, Shay Totten, and Seth Zuckerman.

Organizations: HKH Foundation, Rockefeller Family Fund, Rockefeller Brothers Fund, CS Foundation, Tides Foundation, the Rockridge Institute, and the Climate Protection Campaign.

ABOUT THE
TOMALES BAY INSTITUTE

The Tomales Bay Institute is a network of thinkers and doers who promote public understanding of common wealth. Please visit us at www.onthecommons.org.

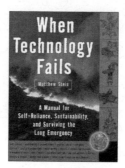